Everyone in Tory's house was happy. "Gran comes this afternoon!" said Tory.

"I will precook our dinner now," said Dad.

Mom said, "I will take Bart to preschool."

At school Tory quickly
started to unpack her books.
Then Ms. Stern spoke about
Show and Tell Day. "You must
plan your speeches carefully,"
she said.

"What will I tell my class
about?" Tory wondered.

After school Tory said,
"When will Gran be here?"

"I know you dislike waiting,"
said Mom with a smile.

When Gran came, Tory
quickly rushed outside to
kiss her.

Tory helped Gran unpack
her bags and boxes. Some
boxes were bigger than others.
Gran gave her a small box to
unwrap.

"This is the softest hat in
the world," Tory said. "What a
lovely surprise!"

"Tory, this is your surprise," Gran said with tenderness. Gran took out an old book with a little rusty lock. She gently unlocked it.

Then Gran talked about all the photos in the album.

Later, Tory told Gran about
Show and Tell Day. "I don't
have anything to talk about,"
Tory said.

"I know you do," said Gran
with a twinkle in her eye.

Gran gently handed her the
photo album. "Please use this
for Show and Tell," Gran said.

As she took the book, Tory
was speechless.

At school Tory's friends
said, "What a cool book!"

"I bet it's the oldest book in
our school," said Ms. Stern.
"Tell us about it, Tory."

Tory's speech was one of
her proudest moments.

The End